Faithful Flocks

Scriptures and Prayers for Shepherds

Sallie Dawkins

Faithful Flocks: Scriptures and Prayers for Shepherds

Library of Congress Control Number: 2025923362

eBook ISBN 978-1-955861-65-6; Print Paperback ISBN 978-1-955861-63-2

Published by
Open Ark Books
and Firebrand United, LLC
216 Skywatch Dr. PMB 206
Danville, Kentucky
40422 USA

Author Website:
www.SallieDawkins.com

Original text and photographs by Sallie Dawkins. Image conversion to storybook effect using ChatGPT5.

For Andrew

Your gentle heart and unwavering compassion inspire me every day.

Andrew with newborn lamb ♡

6792

"The Lord
is good
to everyone.
He shows his
mercy to
everything
He made."

Psalm 145:9

Creator of all,
we honor Your ground,
in flowers and trees,
Your love is found.
May we cherish
the animals
and friends we share,
and nurture Your world
with gratitude
and prayer.

"So God made
every kind
of animal.
He made the
wild animals,
the tame animals,
and all the small
crawling things.
And God saw that
this was good."

Genesis 1:25

Dear God,
thank You for
the beauty of creation
around us.
Open our ears to hear
the animals and learn
from the world
You have made.
We are grateful
for Your love.
Please guide us
to be more like
these little ones-
curious, joyful, and
trusting in Your hands.
Amen.

"Learn all you can
about your sheep.
Take care of
your goats the
best you can."

Proverbs 27:23

Thank You, God,
for all our sheep,
for hills to climb
and grass to keep.
Bless our work
from dawn till night,
and keep our flock
safe in Your light.
Amen.

"I own all the
animals on a
thousand hills.
I know every bird
in the mountains.
Everything that
moves in the fields
is mine."

Psalm 50:10-11

Thank You, God,
for the creatures
we see,
the cows, the goats,
the buzzing bee.

Help us love them
and treat them right.
Caring for them
is a pure delight!

"My sheep listen
to My voice.
I know them,
and they
follow Me."

John 10:27

Heavenly Father,
help us to see Your
goodness in every part
of creation.
Strengthen us
to protect the earth
and its creatures
so they can flourish.
Amen.

"The shepherds
went back
to their sheep,
praising God and
thanking Him
for everything
they had seen
and heard.
It was just as
the angel
had told them."

Luke 2:20

Thank You, Lord,
for the sky so blue,
for butterflies
and birds to view.

Help us see the beauty
in all we find,
reflecting Your love,
both gentle and kind.

"They will be like sheep that eat and lie down in peace- and no one will bother them."

Zephaniah 3:13

Shining sun,
so warm and bright,
thank You, God,
for Your guiding light.
May it shine
over the flocks
and fields we keep,
bringing joy
and watchful peace.

"We are
Your people,
the sheep
of Your flock.
We will praise
You forever.
We will praise You
forever
and ever!"

Psalm 79:13

May the shepherd's heart
be brave and true,
guided by Your strength
in all he must do.

When lambs stray off
or shadows grow deep,
let his faith be the light
that guides his sheep.

"You open Your
hands and give
every living thing
all that it needs."

Psalm 145:16

Each season unique,
from winter to fall,
with Your grace, Lord,
we welcome them all.

May the cycles of life
remind us each day,
of Your steadfast love,
come what may.

"God has shown you His grace in many different ways. So be good servants and use whatever gift He has given you in a way that will best serve each other."

1 Peter 4:10

May we pause, Lord,
to see Your grace,
in every field
and quiet place.
Teach us to care
for flock and land,
guided always
by Your loving hand.

"Let the peace
that Christ gives
control your
thinking.
It is for peace
that you were
chosen to be
together in one
body. And always
be thankful."

Colossians 3:15

Heavenly Father,
thank You for
the gift of this earth,
for the vibrant fields
that bring life
and worth.

We pray we may honor
Your creation
with care,
loving every creature,
beyond compare.

"But ask the animals,
and they will
teach you.
Or ask the
birds of the air,
and they will
tell you."

Job 12:7

Let faith guide us to
care for God's creation
and share His gifts
joyfully.

God, we are grateful
for Your daily blessings.

Your generosity makes
our hearts sing with joy!

For every person,
animal, and plant,
we offer our gratitude.

"Look at the birds.
They don't plant,
harvest, or save
food in barns, but
your heavenly
Father feeds them.
Don't you know
you are worth
much more than
they are?"

Matthew 6:26

When storms may come
and skies turn gray,
thank You, God, for
guiding our way.

Fill us with courage to
face the unknown.
With faith in You,
we're never alone.

"You have
put gladness
in my heart."

Psalm 4:7 WEB

Dear God,
thank You
for the amazing world
full of animals
You have created.

Help us cherish
and take care of them,
just like You
take care of us.

We are grateful
for all that You provide
and for the lessons
we learn from
Your creation.
Amen.

"People in all the
towns of Judah
will live together
in peace.
Farmers and those
who move around
with their flocks
will live peacefully
together in Judah."

Jeremiah 31:24

Heavenly Father,
thank You
for the animals
that teach us
about Your ways.
Help us listen to them
as we learn to
care for
Your creation.
Amen.

Then a voice came
from the throne
and said,
"Praise our God,
all you who
serve him!
Praise our God,
all you small
and great who
honor him!"

Revelation 19:5

USA

May we always
treat others
with kindness and love,
just as
we would wish
to be treated.

"Now, our God,
we thank You,
and we praise
Your glorious
name!"

1 Chronicles 29:13

Dear God,
teach us to lead with
loving hearts,
to care for others
and do our part.
Fill us with joy
as we obey,
and follow You
in every way.
Amen.

"The LORD is
my shepherd.
I will always
have everything
I need."

Psalm 23:1

Dear God,
bless the shepherds
with strength and care,
for guiding the flock,
they're always there.

May they find pasture,
safety, and peace.
Your love and protection
shall never cease.
Amen.

"The Lamb
in front of the
throne will be their
Shepherd.
He will lead them
to springs
of water
that give life.
And God
will wipe away
every tear
from their eyes."

Revelation 7:17

Lord,
You made
the sheep,
the hens,
and me!
Thank You
for letting
us care
for all
You've made
so lovingly.
Amen.

"You were
like sheep
that went the
wrong way.
But now
you have
come back
to the Shepherd
and Protector
of your lives."

1 Peter 2:25

Lord, we praise You
with joyful hearts!
Just like the shepherds
long ago,
help us notice Your
goodness
and share Your love
wherever we go.
Amen.

God blessed them
and said to them,
"Have many children.
Fill the earth and
take control of it.
Rule over the fish
in the sea and the
birds in the air.
Rule over every
living thing that
moves on the earth."

Genesis 1:28

Dear Lord,
help us always
to be kind
and caring
toward animals,
Just as You
care for them.
May we learn
to be gentle
and responsible!
Amen.

"Good people
take good care
of their animals,
but the wicked
know only how
to be cruel."

Proverbs 12:10

Dear God above, so good
and true, We thank You
for the work You do.
For sheep and hens and
cows that play,
we're thankful for them
every day.
Please help us care
for all we see,
with gentle hearts and
love like Thee.
Remind us, Lord,
we're in Your keep-
forever loved,
we are Your sheep.
Amen.

EXTENDING THE STORY

PRAY TOGETHER: Say short prayers for pets, farm animals, or wild animals.

HANDS-ON LEARNING: Try drawing, listening for birds, or caring for pets.

MEMORY VERSE: Pick a scripture verse and practice it all week.

DISCUSSION STARTER: Talk about how God is our Good Shepherd. Ask: "What can we learn from animals about God's love?"

NEXT STEPS

The illustrations for this story were created from photos taken from the author's time on a real farm.

If you enjoyed this story, please consider leaving an honest review on Amazon or Goodreads. Your words have the power to inspire other readers, and spread the important message of love and support this book carries.

To connect with the author, visit www.SallieDawkins.com.

Made in the USA
Monee, IL
05 November 2026

67378080R00059